Smart City Transportation

Leveraging Technology for Efficient Traffic Management and Passenger Information

Kirby Thornecroft

Table of Contents

1 Introduction to Smart City Transportation . . . 5
1.1 Defining Smart City Mobility . . . 9
1.2 Technological Advancements . . . 13

2 Intelligent Transportation Systems (ITS) 17
2.1 Adaptive Traffic Control . . . 21
2.2 Traffic Monitoring & Analysis . . . 24
2.3 Incident Management Systems . . . 27

3 Real-Time Passenger Information Systems . . . 31
3.1 Data Acquisition & Processing . . . 34
3.2 Multimodal Journey Planning . . . 37
3.3 User Interfaces & Accessibility . . . 39
3.4 Data Visualization & Presentation . . . 43
3.5 System Integration & Interoperability . . . 46

4 Data Analytics for Transportation . . . 51
4.1 Predictive Modeling . . . 54
4.2 Performance Measurement . . . 58

5 Emerging Technologies in Transportation . . . 63
5.1 Autonomous Vehicles . . . 66
5.2 Connected Cars . . . 69
5.3 Shared Mobility Services . . . 72

6 Smart City Transportation Planning & Policy . . . 77
6.1 Sustainable Transportation Strategies . . . 80
6.2 Public-Private Partnerships . . . 83
6.3 Policy & Regulatory Frameworks . . . 85
6.4 Stakeholder Engagement . . . 89
6.5 Future Trends & Challenges . . . 92

7 Case Studies & Best Practices . . . 97
7.1 City A: Successful ITS Implementation . . . 101
7.2 City B: Innovative Passenger Info System . . . 104

1 Introduction to Smart City Transportation

Smart city transportation represents a paradigm shift from traditional approaches. It's not merely about technological upgrades; it's a fundamental restructuring of how we design, manage, and experience urban mobility. This involves a holistic integration of various technologies and strategies to optimize efficiency, enhance safety, and improve the overall quality of life for city dwellers. We're talking about a complete transformation, moving beyond reactive measures to a proactive, data-driven model that anticipates and addresses challenges before they arise. This proactive approach necessitates a strong foundation in technological innovation and sophisticated data analysis capabilities.

The core of smart city transportation lies in its technological backbone. Advanced sensor networks embedded throughout the city gather real-time data on traffic flow, vehicle speeds, pedestrian movements, and even air quality. This data fuels the intelligent transportation systems

(ITS) discussed in the next chapter, enabling dynamic adjustments to traffic signals, efficient routing of emergency vehicles, and proactive identification of potential congestion points. High-speed communication networks, such as 5G, are crucial for the seamless transmission of this vast amount of information, ensuring that systems operate with minimal latency and maximum responsiveness. Powerful data analytics platforms then process this information, identifying patterns, predicting future scenarios, and generating insights that inform policy decisions and operational strategies. This is far beyond simple traffic management; it's a system of interconnected networks, constantly learning and adapting.

Effective implementation hinges on a multi-faceted approach encompassing robust data management strategies. Real-time data streams from diverse sources—GPS trackers on buses, traffic cameras, smart parking sensors, mobile phone location data—require sophisticated processing and integration. Data quality and reliability are paramount; inaccuracies can lead to flawed decisions and operational inefficiencies. Therefore, rigorous data validation and cleansing procedures are essential. This also means establishing clear data governance frameworks, defining access control measures, and ensuring compliance with data privacy regulations. A well-structured data management system is the very foundation on which the entire smart city transportation system rests. It's not just about collecting data; it's about effectively managing, analyzing, and using that data to make impactful decisions. Success also depends on seamless integration with existing infrastructure and services. Retrofitting existing systems with smart technology often presents significant challenges. Compatibility issues between different platforms and systems are a major concern. Planning for interoperability from the outset is vital. The aim is not to replace existing systems entirely but to enhance them. We need a phased approach to upgrading infrastructure, ensuring that new smart technologies can smoothly coexist and interact with pre-existing systems. This integrated approach requires meticulous coordination between various stakeholders – government agencies, transportation providers, technology vendors, and, crucially, the citizens themselves.

Furthermore, the human element is crucial. Smart city transportation isn't just about technology; it's about people. Designing user-friendly interfaces for real-time information systems is critical for effective communication and widespread adoption. Accessibility is a key consideration; systems should be easily navigable by all demographics, including those with disabilities. Clear, concise, and readily understandable information is crucial for enhancing passenger experience and improving decision-making. This involves considering the needs of various user groups – commuters, tourists, delivery drivers – and designing systems that cater to their specific requirements. Engagement with citizens is vital to ensure their needs are met and that the technology serves them effectively.

A successful smart city transportation strategy must prioritize sustainability. Reducing carbon emissions is a major goal, and smart transportation plays a significant role in achieving this. Optimized traffic flow reduces fuel consumption, while promoting the use of public transport, cycling, and walking contributes to a greener city. Integration with other sustainable initiatives, such as renewable energy sources, helps to minimize the environmental impact of the entire transportation system. A sustainable approach also necessitates careful consideration of the lifecycle costs of technologies, ensuring that smart systems are both efficient and environmentally responsible. The long-term sustainability of the system relies not just on immediate technological advancements, but on the consideration of long-term environmental impacts.

Finally, the economic aspects of smart city transportation are significant. Investing in smart technologies requires careful planning and effective resource allocation. Public-private partnerships can leverage expertise and funding from both the public and private sectors, ensuring the feasibility of large-scale projects. Thorough cost-benefit analyses are essential to demonstrate the long-term economic viability of such initiatives. A strong economic model ensures that investments are justified and lead to tangible improvements in the efficiency and sustainability of the city's transportation network. This requires transparency and accountability, assuring the public that resources are being used effectively and delivering substantial returns.

1.1 Defining Smart City Mobility

Smart city mobility transcends the simple movement of people and goods; it's a complex interplay of technology, infrastructure, and human behavior, aiming for an efficient, sustainable, and equitable transportation system. Unlike traditional approaches focused on individual modes, smart city mobility embraces integration and optimization across all transport options – from public transit and cycling to ride-sharing and autonomous vehicles. It's about creating a holistic network where travel choices are informed, seamless, and user-friendly. This necessitates a fundamental shift from reactive to proactive management, relying on real-time data and predictive analytics to anticipate and address challenges before they arise.

The core of smart city mobility rests on the ability to collect, analyze, and utilize massive amounts of data. This data, sourced from various sensors embedded within the infrastructure and vehicles, provides granular insights into traffic patterns, passenger flows, and environmental conditions. Sophisticated algorithms then process this information, informing decisions related to traffic signal optimization, route planning, and resource allocation. The ultimate goal is to reduce congestion, improve safety, minimize environmental impact, and enhance the overall travel experience for all citizens. Think of it as a dynamic, self-regulating system constantly adapting to real-time needs.

This isn't just about technological advancements; successful implementation hinges on comprehensive planning and policy. Effective smart city mobility requires strong collaboration between government agencies, private sector companies, and citizens themselves. Open data initiatives play a crucial role, allowing developers and researchers to access and leverage transportation data to create innovative applications and services, fostering a vibrant ecosystem of mobility solutions. Furthermore, robust regulatory frameworks are essential to ensure data privacy, system security, and equitable access for all members of the community.

Consider, for instance, the potential impact on vulnerable populations. Smart city mobility solutions should be designed with accessibility and inclusivity at their heart, ensuring that elderly individuals, people with disabilities, and low-income residents can equally benefit from improved transportation options. This means designing user interfaces that are intuitive and accessible, integrating real-time information into assistive technologies, and promoting affordable and convenient transport alternatives. Ignoring these considerations would result in a system that exacerbates existing inequalities, contradicting the very principles of a smart city.

The financial implications are substantial, requiring significant investment in infrastructure upgrades, technological advancements, and ongoing maintenance. Public-private partnerships can play a vital role in securing funding and expertise, combining the financial resources of the private sector with the regulatory oversight and public interest focus of government agencies. Carefully constructed financial models, which consider both the short-term and long-term costs and benefits, are essential for ensuring the financial viability and sustainability of smart city mobility initiatives. It's not just about deploying technology; it's about creating a sustainable, financially sound system that benefits the entire community.

Moreover, smart city mobility is inextricably linked to broader sustainability goals. By optimizing traffic flow and encouraging the use of public transport and active travel modes such as walking and cycling, smart city initiatives contribute to reduced greenhouse gas emissions and improved air quality. The integration of electric vehicles and alternative fuels further enhances environmental benefits. Ultimately, smart city mobility should strive towards a future where transportation is efficient, sustainable, and contributes positively to the overall health and well-being of the urban population, fostering a greener and healthier city environment. Sustainable practices are not merely an add-on; they are fundamental to the concept's success.

Successful implementation also demands a focus on data security and privacy. The vast amounts of data collected as part of smart city mobility initiatives must be handled responsibly, ensuring the protection of personal information and preventing misuse. Robust cybersecurity measures are crucial to safeguard against potential attacks and data breaches. Clear data governance policies, combined with transparent data sharing agreements, are essential to building public trust and ensuring the ethical and responsible use of data. Openness and transparency are not optional; they are vital for public acceptance and successful long-term implementation. Failing to address these concerns could lead to public mistrust and the failure of the entire initiative.

Finally, the ongoing evolution of technology necessitates a flexible and adaptable approach to smart city mobility planning. Emerging technologies, such as autonomous vehicles and drone delivery services, will continue to reshape the urban transportation landscape. Long-term planning needs to anticipate and integrate these advancements, ensuring that the smart city mobility system is future-proof and capable of adapting to technological changes and evolving societal needs. Continuous monitoring, evaluation, and adaptation are key to ensuring that the system remains effective and relevant in the long term. The challenge lies in remaining proactive and flexible in the face of constant technological advancements.

1.2 Technological Advancements

The foundation of any successful smart city transportation system lies in its technological backbone. This isn't simply about installing new gadgets; it's about a holistic integration of diverse technologies working seamlessly together. Think of it as a sophisticated orchestra, where each instrument – sensors, communication networks, data analytics platforms – plays its part to create a harmonious, efficient whole. The effectiveness of the entire system depends on the quality and reliability of each individual component and their interaction.

Consider the pervasive deployment of sensors. These aren't just your typical traffic cameras. We're talking about a network of sophisticated sensors embedded throughout the city's infrastructure, constantly collecting a vast amount of data. From vehicle counts and speeds to environmental conditions and even pedestrian movements, this data forms the raw material for intelligent decision-making. The accuracy and density of this sensor network directly impact the precision and timeliness of traffic management strategies. High-resolution sensors provide a far richer data set, leading to more refined analysis and better-informed interventions. Furthermore, the reliability of these sensors, their resistance to damage or interference, is crucial for maintaining the integrity of the whole system.

Communication networks are the vital arteries of this

system, transmitting the raw data collected by the sensors to central processing units. The speed and capacity of these networks determine the responsiveness of the system. Real-time traffic management demands near-instantaneous data transmission to allow for dynamic adjustments to traffic signals and other controls. The choice of communication technology, whether it's cellular networks, Wi-Fi, or dedicated short-range communication systems, significantly influences the efficiency and effectiveness of the entire infrastructure. Robust and redundant communication channels are crucial to ensure uninterrupted data flow, even in the event of network failures or unexpected surges in demand. This redundancy is often overlooked but represents a critical factor in the overall system's resilience and reliability.

Data analytics sits at the heart of this technological ecosystem. It's not enough simply to collect data; we need to transform it into actionable insights. Sophisticated algorithms process the massive data streams, identifying patterns, predicting congestion hotspots, and optimizing traffic flow. These analytical capabilities inform decisions about signal timing, routing, and resource allocation, leading to significant improvements in efficiency and reduced congestion. This process isn't static; these algorithms are constantly learning and adapting, improving their predictive capabilities over time as they process more data. The development and implementation of these analytical capabilities need to take into account not only

the immediate needs but also long-term city planning.
Beyond the core elements, advancements in other areas are crucial. High-precision GPS technology, for example, enables accurate tracking of vehicles and improves the accuracy of real-time traffic information. The increasing sophistication of mobile applications allows for seamless integration of passenger information systems with other modes of transportation, providing commuters with comprehensive, real-time journey planning tools. These features are no longer mere luxuries; they're essential components of a truly smart city transportation system, enhancing usability and improving the overall passenger experience. The importance of user-friendly and accessible interfaces for all members of society cannot be overstated, ensuring inclusivity in the use of these sophisticated technologies.

The ongoing evolution of technological advancements holds immense potential for further enhancing smart city transportation. The integration of artificial intelligence and machine learning promises even more refined predictive models and adaptive traffic management strategies. The emergence of new communication technologies and sensor networks will continue to improve data quality and transmission speeds. These ongoing developments must be carefully integrated into existing systems, ensuring seamless compatibility and optimal performance. Failure to do so could lead to systems that are inefficient and difficult to manage, not delivering on the promise of a truly

smart and efficient urban environment. Therefore, continuous monitoring and upgrades are vital for maintaining a competitive edge in utilizing the power of technological advancements for a better urban transportation system.
Finally, the successful implementation of these technologies requires not just technological expertise but also careful planning, effective coordination, and robust collaboration amongst stakeholders. This includes city planners, transportation authorities, technology providers, and the public itself. Smart city transportation is not just about technology; it's about people. It is the effective integration and proper utilization of all these components that will ultimately determine the success of smart city transportation initiatives. Only through a thoughtful, comprehensive approach can we truly harness the power of technology to create a more efficient, sustainable, and user-friendly urban transportation experience.

2 Intelligent Transportation Systems (ITS)

Optimizing traffic flow is paramount in any smart city initiative. Adaptive traffic control systems, the backbone of Intelligent Transportation Systems (ITS), achieve this by dynamically adjusting signal timings based on real-time traffic conditions. These systems utilize a network of sensors embedded throughout the road network, constantly monitoring traffic volume, speed, and density. Sophisticated algorithms analyze this data, identifying congestion hotspots and predicting future traffic patterns. Consequently, signal timings are adjusted to prioritize traffic flow where it's needed most, minimizing delays and improving overall network efficiency. This dynamic approach contrasts sharply with traditional fixed-time signal control, which often leads to unnecessary stops and wasted fuel. The efficiency gains are significant; studies have shown that adaptive traffic control can reduce delays by up to 25%, depending on the specific implementation and urban context.

Traffic monitoring and analysis forms a crucial layer within ITS. This involves not just collecting data but interpreting it to gain actionable insights. Advanced sensor technologies, such as radar, lidar, and video cameras, provide a comprehensive view of traffic conditions. Data fusion techniques combine information from multiple sources, creating a more accurate and holistic picture. This rich data set then undergoes rigorous analysis, employing statistical methods, machine learning algorithms, and data visualization tools to pinpoint recurring congestion patterns, identify bottlenecks, and assess the effectiveness of existing traffic management strategies. Understanding these patterns allows city planners to make data-driven decisions, focusing infrastructure investments where they'll have the most impact. Imagine pinpointing a specific intersection consistently causing delays at rush hour—this data enables targeted interventions, like adding lanes, optimizing signal timing, or improving pedestrian crossings.

Incident management is another vital component. Real-time incident detection systems are crucial for rapid response to accidents, breakdowns, or other unexpected events. This typically involves integrating data from various sources, including emergency services, traffic cameras, and citizen reports through mobile apps. When an incident occurs, the system automatically alerts relevant authorities, providing them with precise location information, traffic conditions, and potential impact. This swift response helps minimize disruption, reduce congestion buildup, and improves public safety. A well-designed incident management system can significantly decrease response times, improving emergency response and leading to a safer, more efficient transport network. Imagine the difference between a 20-minute response time compared to a 5-minute one in a critical situation; it's the difference between potentially life-saving intervention and a prolonged period of danger and gridlock.

Beyond the immediate benefits of reduced congestion and improved safety, ITS offers significant long-term advantages. Data gathered by these systems provides valuable insights into transportation patterns and infrastructure needs. This informs long-term planning decisions, leading to more sustainable and efficient transportation networks. By constantly monitoring and analyzing traffic data, city planners can identify areas needing improvements, prioritize infrastructure projects, and adapt to evolving travel patterns. This approach ensures investment dollars are used effectively, creating a truly intelligent and resilient transportation system. For instance, identifying consistent peak-hour congestion on a specific route might reveal the need for a new public transportation option or expanded road capacity, ultimately benefiting commuters and city life.

Furthermore, the data collected by ITS goes beyond traffic flow. Information on vehicle emissions, fuel consumption, and travel times offers essential insights for environmental management and sustainability initiatives. This allows cities to implement strategies aimed at reducing carbon footprint, improving air quality, and promoting eco-friendly transportation modes. By understanding the environmental impact of various transportation choices, policy-makers can create incentives for using public transport, cycling, or walking, helping to build a greener and healthier urban environment. The data allows the quantification of the impact of certain initiatives, allowing for evidence-based policy-making that goes beyond anecdotal observations.

Finally, the success of ITS implementation hinges on effective data sharing and collaboration. Different agencies and stakeholders, including traffic management departments, public transport operators, and emergency services, must work together to create a unified and integrated system. Data exchange protocols, interoperability standards, and shared data platforms are key to achieving seamless integration. The free flow of information helps to prevent redundancy, ensures consistency, and allows for a truly holistic approach to transportation management. A coordinated effort is essential to ensure that all parts of the system work in harmony, leading to a smoother and more reliable experience for all road users. The result is a much more efficient and reactive transportation system.

2.1 Adaptive Traffic Control

Adaptive traffic control systems represent a significant advancement in managing urban traffic flow. They dynamically adjust traffic signal timings based on real-time traffic conditions, aiming to reduce congestion, improve journey times, and enhance overall network efficiency. Unlike traditional fixed-time signal control, which operates on pre-programmed schedules regardless of actual traffic demand, adaptive systems constantly monitor traffic patterns, using a variety of sensors and data sources. This allows the system to prioritize traffic flow on the most congested streets, optimizing signal timing to alleviate bottlenecks and reduce delays.

The core of an adaptive traffic control system lies in its ability to process vast amounts of data. Sensors embedded within the road network, such as inductive loop detectors, video cameras, and radar systems, collect real-time information on traffic volume, speed, and density. This data is transmitted to a central control system, typically utilizing a sophisticated communication network. The central system then employs algorithms to analyze the collected data and determine the optimal signal timing for each intersection. These algorithms often incorporate advanced techniques from operations research and artificial intelligence to optimize traffic flow across the entire network, rather than just individual intersections.

Several different algorithmic approaches exist for adaptive traffic control. Some methods focus on minimizing total network delay, while others prioritize minimizing the number of stopped vehicles or reducing the overall travel time across the network. The choice of algorithm depends on the specific goals of the traffic management system and the characteristics of the road network. Furthermore, the effectiveness of any algorithm can be significantly impacted by factors such as the accuracy of the data collected by the sensors, the communication network's reliability, and the computational power available to the control system. A robust system must account for these variables and dynamically adapt to changing conditions.

Implementing adaptive traffic control requires careful planning and consideration. First, a comprehensive assessment of the road network is crucial to identify critical intersections and potential bottlenecks. This involves detailed traffic modeling and analysis, often using specialized simulation software to forecast traffic flow under different control strategies. Once the optimal sensor placement and communication infrastructure are determined, the system can be implemented and tested. This phase often involves extensive field testing and fine-tuning of the control algorithms to ensure optimal performance under real-world conditions. Ongoing monitoring and adjustments are essential, adapting to variations in traffic patterns throughout the day, week, and across different seasons. Feedback mechanisms, incorporating data from user experiences and further analysis, allow for continu-

ous improvement in the system's performance.
Beyond simply optimizing signal timing, advanced adaptive traffic control systems can integrate with other smart city initiatives. For example, real-time information from incident management systems, such as reports of accidents or road closures, can be incorporated into the control algorithms to reroute traffic and minimize the impact of disruptions. Similarly, integration with public transportation systems can allow for the prioritization of buses and other forms of public transport, promoting the use of sustainable modes of travel. Data visualization tools can present this traffic information to drivers, offering them alternative routes and reducing the overall strain on the network. Furthermore, by predicting traffic congestion based on historical data and current trends, adaptive systems can proactively manage traffic flow, mitigating potential problems before they escalate into major delays. These proactive measures contribute not only to smoother traffic but also to reducing fuel consumption, lowering emissions, and enhancing the overall quality of life in the smart city. The potential benefits of implementing such sophisticated systems extend to a broader environmental and economic impact, representing a crucial component of sustainable urban development.

2.2 Traffic Monitoring & Analysis

Effective traffic monitoring and analysis forms the cornerstone of any successful smart city transportation system. It's not merely about knowing how many cars are on a road; it's about understanding the why behind traffic patterns, anticipating congestion, and proactively managing flow. This understanding enables optimized signal timing, efficient emergency response, and ultimately, a smoother, safer commute for everyone.

Real-time data acquisition is crucial. Sensors embedded in roadways, strategically positioned cameras, and GPS data from vehicles provide a continuous stream of information on traffic speed, volume, and density. This data, however, is only as useful as its processing and interpretation. Sophisticated algorithms are necessary to filter noise, identify anomalies, and translate raw data into actionable insights. For example, a sudden drop in speed in a specific area might indicate an accident, triggering an automated alert to emergency services. Accurate and timely incident detection minimizes delays and enhances public safety.

Data visualization is paramount in making this complex information accessible and understandable. Interactive maps displaying real-time traffic flow, heatmaps highlighting congestion hotspots, and animated visualizations of traffic patterns offer a clear picture of the transportation network's current state. These tools enable transportation managers to quickly grasp the situation, identify critical areas, and implement appropriate countermeasures. Think of it like a doctor using X-rays to diagnose a patient – the data is raw, but the visualization makes the diagnosis clear. Such immediate visibility significantly improves response times to incidents and allows for more effective resource allocation.

Beyond simply reacting to immediate events, effective traffic analysis involves predictive modeling. By analyzing historical traffic data, weather patterns, and even calendar events (like concerts or sporting events), we can anticipate future congestion patterns. This predictive capability allows for proactive interventions, such as adjusting signal timing in advance of anticipated congestion or advising commuters on alternative routes. Predictive modeling isn't about crystal balls; it's about using data-driven insights to anticipate likely scenarios and prepare accordingly, ultimately reducing overall travel time and fuel consumption. However, data is only part of the equation. The success of any traffic monitoring and analysis system hinges on the quality of the data collected and the robustness of the analytical techniques employed. Regular calibration and maintenance of sensors are vital to ensure accuracy. The algorithms used must be regularly updated and refined to adapt to evolving traffic patterns and technological advancements. This is an ongoing process of improvement, refinement, and adaptation, not a one-time solution. It requires constant monitoring and evaluation, utilizing feed-

back loops to ensure optimal performance.
Furthermore, effective implementation necessitates a holistic approach. Data sharing between different agencies and organizations—from traffic management centers to public transportation authorities—is crucial. A unified platform allowing access to comprehensive data from various sources allows for a comprehensive understanding of the entire transportation system, rather than isolated pockets of information. This comprehensive view enables a more coordinated and effective approach to traffic management. Effective collaboration is, therefore, not just a suggestion, but a fundamental prerequisite.
Finally, the ethical considerations surrounding data privacy must not be overlooked. While collecting and analyzing traffic data offers substantial benefits, it is crucial to ensure that individual privacy is protected. Data anonymization techniques, robust security measures, and transparent data governance policies are vital to build public trust and ensure the responsible use of this valuable information. Balancing the benefits of improved traffic management with the protection of individual rights is a crucial aspect of successful implementation. Failing to address this can lead to public distrust and hinder the overall effectiveness of the system. Open communication and transparency are key to establishing public confidence in the system.

2.3 Incident Management Systems

Efficient incident management is crucial for a smoothly functioning smart city transportation system. Delays caused by accidents, road closures, or unexpected events ripple outwards, impacting commutes, public transit schedules, and overall city efficiency. A robust system needs to detect incidents quickly, accurately assess their impact, and deploy resources effectively for resolution. This requires a comprehensive approach, integrating various technologies and strategies.

Real-time data acquisition is paramount. Sensors embedded in roads, cameras at intersections, and GPS data from vehicles provide continuous streams of information. This data needs to be processed and analyzed in real-time, identifying deviations from normal traffic patterns – sudden slowdowns, unexpected stops, or unusual traffic volumes – that may indicate an incident. Smart algorithms can analyze this data, filtering out noise and prioritizing potential incidents based on severity and impact. Automated alerts can then be triggered, notifying relevant authorities.

Effective communication is the backbone of incident management. A centralized platform is necessary, enabling seamless information exchange between various stakeholders. This includes traffic management centers, emergency services, public transportation operators, and even the general public. The system should have the capacity to send out targeted alerts to affected drivers and commuters, providing real-time updates on road closures, alternative routes, and estimated delays. Clear, concise messaging is vital; ambiguous or confusing information can worsen the situation.

The efficient deployment of resources is another critical aspect. Once an incident is identified, the system should automatically dispatch appropriate resources. This could include tow trucks, emergency medical services, or traffic police, depending on the nature of the incident. Optimizing resource allocation based on real-time data ensures that help arrives quickly and efficiently, minimizing disruption. This requires sophisticated routing algorithms that consider traffic conditions and other factors to determine the quickest routes for emergency vehicles.

Integrating incident management with other smart city systems enhances overall efficiency. For example, real-time data from incident management systems can feed into adaptive traffic control systems, allowing for dynamic adjustments to signal timing to alleviate congestion around affected areas. Similarly, the system can be integrated with passenger information systems, providing commuters with up-to-the-minute updates on delays and alternative travel options. This interconnectedness ensures a coordinated response to incidents, minimizing their overall impact.

However, the effectiveness of an incident management system hinges not only on technology but also on people. Training and clear protocols are vital for all involved personnel. Effective communication channels and well-defined roles and responsibilities are essential for ensuring coordinated responses. Regular drills and simulations allow for testing and improving response procedures, identifying potential weaknesses and strengthening overall system resilience.

Furthermore, a robust feedback mechanism is important for continuous improvement. Post-incident analyses can help identify areas for improvement in the system. This could involve refining algorithms, adjusting protocols, or enhancing training. Collecting user feedback through surveys or online platforms provides valuable insights into the effectiveness of the system from a user's perspective. This iterative process is crucial for continuously optimizing the incident management system and adapting to evolving needs.

Finally, consider data privacy and security. The data collected by the system contains sensitive information, and its security must be a paramount concern. Implementing robust security measures, including encryption and access control, is vital to prevent unauthorized access or data breaches. Compliance with relevant privacy regulations is also crucial, ensuring the ethical handling of user data. The system's design should carefully balance the need for efficient incident management with the protection of personal information. Transparency and accountability in data handling build public trust, a crucial element for the success of any smart city initiative.

3 Real-Time Passenger Information Systems

Real-time passenger information systems are crucial for efficient and user-friendly public transport. Effective systems rely on a robust foundation of data acquisition and processing, ensuring accurate and timely information delivery. This involves integrating data from various sources such as GPS tracking on vehicles, smart card ticketing systems, and even social media feeds reporting disruptions. Sophisticated algorithms process this raw data, cleaning it, filtering out errors, and transforming it into easily understandable formats for display. The challenge lies not only in the sheer volume of data but also in ensuring data quality and consistency across different sources, a task that requires careful planning and ongoing maintenance. Efficient data management strategies are paramount to a successful system.

Multimodal journey planning is a key feature of modern real-time passenger information systems. Users expect seamless integration across different transport modes –

buses, trains, subways, trams, even ride-sharing services. A well-designed system provides not just information on the next available bus, but an optimal route considering various factors like travel time, cost, and accessibility preferences. This requires advanced algorithms that can search and optimize across complex networks and adapt in real-time to changing conditions, such as delays or unexpected disruptions. Such planning tools are incredibly powerful, facilitating informed travel decisions and enhancing overall travel experience. The development of a user-friendly, intuitive interface is also essential for its effectiveness.

The user interface is the face of the entire system. It must be intuitive, accessible, and informative, catering to a diverse population with varying levels of technological proficiency. This demands a careful consideration of design principles, such as clear visual hierarchy, easily understandable icons, and multiple language support. Accessibility is paramount; the system must be usable by individuals with visual, auditory, or motor impairments, adhering to accessibility standards like WCAG. This includes features such as screen readers compatibility, alternative text descriptions for images, and clear audio announcements at stations. Designing for inclusivity is not merely a matter of compliance; it's about creating a system truly beneficial to everyone.

Effective data visualization and presentation are essential for conveying complex information clearly and concisely. Simple, intuitive displays are more effective than cluttered, data-heavy interfaces. Real-time maps showing vehicle locations, estimated arrival times, and potential delays are extremely useful. Interactive dashboards can provide users with more detailed information, such as historical travel times or service alerts. The challenge is to balance the detail needed for informed decision-making with the simplicity required for a quick and easy understanding of the information. The optimal visualization method will depend on the context and the users' needs.

System integration and interoperability are critical for a successful real-time passenger information system. The system must seamlessly integrate with other transportation systems and services within the city. This includes coordinating with traffic management systems to provide accurate estimates of travel times and displaying real-time traffic conditions on the map. Interoperability between different transport providers is also essential, allowing users to plan journeys involving multiple modes of transport with ease. Achieving this requires strong collaboration and standardized data formats to ensure seamless data exchange between different systems. The benefits of such integration are immense; it creates a truly cohesive and efficient transport network. Careful consideration must be given to data security, privacy concerns, and the potential for system failures. Robust fail-safe measures are crucial to maintaining system reliability and ensuring that passengers are always provided with the best possible information.

3.1 Data Acquisition & Processing

Data acquisition forms the bedrock of any effective real-time passenger information system. Imagine a city's transportation network as a vast, intricate web; we need to gather data from every node to understand its behavior and inform our passengers effectively. This involves integrating data from diverse sources: GPS trackers on buses and trains, smart card readers recording passenger journeys, CCTV cameras monitoring traffic flow, and even social media feeds reporting incidents. Each data stream presents its own unique challenges; inconsistencies in formats, missing data points, and the sheer volume of information necessitate careful planning and sophisticated data management strategies. Real-time data is crucial; delays can render information obsolete, leading to passenger frustration and inefficient resource allocation.

The process begins with defining clear data requirements. What information do we need to provide accurate and timely updates to passengers? This might include estimated times of arrival, service disruptions, platform information, and alternative routes. Once these requirements are established, the selection of appropriate sensors and data collection technologies becomes paramount. Factors such as cost, accuracy, and ease of integration must be carefully considered. The choice of technology will directly impact the overall efficiency and reliability of the entire passenger information system. Careful planning is paramount here, as upgrading or replacing systems is expensive and disruptive.

Data preprocessing is crucial for ensuring data quality and usability. Raw data often contains noise, errors, and inconsistencies. Cleaning and transforming the data are essential steps to prepare it for analysis and visualization. This involves handling missing values, identifying and correcting errors, and converting data into a consistent format suitable for analysis. Data validation is vital: cross-checking information from multiple sources helps to identify and rectify inconsistencies, improving data reliability. Automated processes using scripting and data cleansing tools can significantly improve efficiency here, reducing the manual effort required and mitigating the risk of human error.

Data integration presents another significant challenge. Data from various sources, often in different formats and with varying levels of granularity, needs to be combined into a unified view. This often requires the use of data integration tools and techniques, such as ETL (Extract, Transform, Load) processes. This is more than just combining datasets; ensuring data consistency and accuracy across different sources requires rigorous procedures to prevent errors and inconsistencies from propagating through the system. For example, bus schedules from different operators might use different identifiers for the same bus stop, necessitating careful mapping and standardization before integration.

Data storage and management are critical aspects of the data lifecycle. Real-time passenger information systems generate enormous volumes of data, requiring efficient storage and retrieval mechanisms. This often involves using database management systems that can handle large volumes of data, such as NoSQL or cloud-based solutions which can provide scalability and redundancy. Careful consideration must be given to data security and privacy. Implementing robust security measures to protect sensitive passenger data is paramount. Data needs to be stored securely and accessed only by authorized personnel.

Finally, effective data governance is essential to ensure the long-term success of the system. This involves establishing clear policies and procedures for data acquisition, processing, storage, and access. This includes defining data quality standards, implementing data validation checks, and establishing procedures for handling data breaches. A well-defined data governance framework ensures data consistency, reliability, and compliance with relevant regulations. Regular audits are vital for verifying the adherence to these guidelines and ensuring the system's continued effectiveness and reliability. Without this ongoing monitoring and refinement, the entire process risks becoming unwieldy and unreliable.

3.2 Multimodal Journey Planning

Effective multimodal journey planning hinges on integrating diverse data sources. Think of it like a conductor orchestrating a symphony of transport options – buses, trains, trams, even ride-sharing services. The system must seamlessly combine real-time information on schedules, delays, and even walking times to provide the user with the optimal route. This isn't merely about providing the quickest option; it's about offering various choices, factoring in passenger preferences like walking distance, cost, and accessibility. Success demands accurate, reliable data feeds updated constantly, a robust algorithm that handles complex routing scenarios, and a user-friendly interface that presents this information clearly and concisely.

Building such a system requires a multi-disciplinary approach. Software developers craft the algorithms and interfaces, while transportation planners provide the underlying network data and understand the practical constraints. Data scientists ensure the accuracy and reliability of real-time information, cleaning and processing the influx of data from various sources. Without their expertise, the information presented would be flawed and useless. This collaboration is vital; a software program without context from transportation planning is pointless; similarly, well-planned transportation routes are irrelevant if users cannot easily access the information.

A key challenge lies in data integration. Different transport providers often use incompatible systems. Harmonizing this disparate data requires significant effort, including establishing common data standards and implementing robust data exchange protocols. Imagine trying to fit square pegs into round holes – that's what happens without standardization. Furthermore, real-time updates necessitate reliable communication networks, capable of handling the high volume of data transmitted between different systems. This means not only strong infrastructure but also considering network redundancy to prevent outages, ensuring consistent service. Failures are costly, both financially and in terms of public trust.

User experience is paramount. The best journey planner is useless if users find it difficult to understand and use. Therefore, the interface should be intuitive and accessible to all, regardless of technical skill or disability. This involves careful consideration of the visual layout, text clarity, and the provision of alternative access options, for example, speech-to-text functionality for visually impaired users. Accessibility goes beyond compliance; it is an essential design principle, ensuring equal opportunity for all citizens to utilize the system.

Beyond the immediate route selection, the system should allow users to personalize their experience. Preferences like avoiding stairs, preferring specific transport modes, or setting specific time constraints should be readily incorporated. These customizations enhance user satisfaction and encourage greater uptake. Moreover, the system must provide not only the best route but also provide alternative options, highlighting trade-offs between speed, cost, and convenience. This helps users make informed decisions, and better meets individual needs.

Finally, regular updates and maintenance are crucial for long-term success. Transport networks are dynamic, with new services and infrastructure constantly emerging. The system must be adaptable, allowing for easy integration of new data sources and the updating of existing ones. This iterative process requires continuous monitoring and feedback from users, enabling developers to identify and address shortcomings and enhance the overall user experience. A static system is a failure waiting to happen in a dynamic world. The iterative process is, therefore, a core element of long-term success. Regular updates, user feedback analysis, and continuous system refinement are what separate a merely functional journey planner from a truly successful, valuable public service.

3.3 User Interfaces & Accessibility

Designing intuitive and accessible user interfaces (UIs) is paramount for the success of any real-time passenger information system. A poorly designed UI, regardless of the sophistication of the underlying data, will frustrate users and render the system ineffective. We need to prioritize simplicity, clarity, and ease of navigation. The information should be readily available, without requiring users to hunt for it. Imagine a visually cluttered interface overloaded with information; it's not user-friendly. Think of the frustration a visually impaired person would feel trying to navigate such a system.

Effective UI design involves careful consideration of visual hierarchy, using size, color, and contrast effectively to guide the user's eye. For example, important information, like the next bus arrival time, should be prominently displayed, while less critical details can be relegated to secondary positions. This is crucial for usability. Consider the use of clear and concise language; technical jargon should be avoided, ensuring that the information is understandable to everyone, regardless of their technological literacy. The importance of clear icons and symbols cannot be overstated, especially for those who may have difficulty reading text. Think about the diverse range of users you're designing for: from elderly individuals to those with visual or cognitive impairments.

Accessibility is not merely an afterthought; it's an integral part of the design process. Compliance with accessibility standards, such as WCAG (Web Content Accessibility Guidelines), is essential. These standards provide guidelines for creating websites and applications that are usable by people with disabilities. This includes providing alternative text for images, ensuring sufficient color contrast, and supporting keyboard navigation. Beyond adhering to these technical standards, the design itself needs to embrace a truly inclusive approach. Imagine the frustration of someone using a screen reader encountering poorly structured content or lack of alternative text. This is not just about compliance; it's about offering a dignified and respectful user experience for all.

The use of multiple modalities – visual, auditory, and tactile – can significantly enhance accessibility. Visual impairments can be addressed through audio cues and haptic feedback (touch-based feedback). Consider incorporating voice guidance for navigation, especially in complex transit environments, providing audio directions and real-time updates to users. Furthermore, the design should be adaptable to various screen sizes and devices. The system should seamlessly work on smartphones, tablets, and desktop computers. This requires responsive design, ensuring the UI adjusts automatically to fit different screen dimensions. Consider the variety of devices used by the public and aim for flexibility.

Furthermore, the system needs to offer a variety of input methods. While touchscreens are becoming increasingly prevalent, it's important to offer support for keyboard navigation and alternative input devices for users with motor impairments. Think about the user experience for someone who is physically unable to use a touchscreen and the importance of keyboard accessibility. This approach is not just about compliance with standards but about the broader goal of building an equitable and user-friendly system for all members of the community. The system should also cater to multiple languages, reflecting the linguistic diversity of the city's population, offering translations that are accurate and culturally appropriate.

Testing with diverse user groups is crucial to identify and address potential accessibility issues. This testing should involve users with a range of abilities and disabilities to ensure the system meets the needs of all members of the community. This inclusive design approach makes the system universally accessible. Gathering feedback from this process will help refine the system design, making it better and more inclusive for everyone. This is crucial for building a truly smart city transport system, one that works seamlessly for the diverse population it serves. The integration of assistive technologies, such as screen readers and alternative input devices, into the system's design is critical. The user interface must be designed to work seamlessly with these technologies, ensuring a positive user experience for individuals who rely on them. For example, the system's navigation should be logical and intuitive, even when using a screen reader.

Regular audits of the system's accessibility are crucial for ongoing improvement. Technology evolves, and the accessibility needs of the population also change. Therefore, a proactive approach is crucial, ensuring that the system remains inclusive and accessible for years to come. This requires a continuous cycle of assessment, feedback, and refinement, reflecting the dynamic nature of accessibility requirements. This ongoing improvement process is important, as what works today may not meet the standards tomorrow. Keeping the system current and relevant requires ongoing vigilance, proactive approaches, and a user-centric design philosophy. Remember, a truly smart city transportation system is one that is accessible and inclusive for everyone.

3.4 Data Visualization & Presentation

Effective data visualization is paramount in conveying complex transportation information clearly and concisely to passengers. Think of it this way: you have mountains of data on bus arrival times, train delays, and traffic congestion – but without a clear visual representation, this data is useless. Imagine trying to decipher a spreadsheet filled with thousands of entries detailing bus routes and arrival times; it's overwhelming. Instead, a simple, well-designed map showing real-time bus locations and estimated arrival times becomes an invaluable tool for efficient passenger navigation.

A well-designed visualization should prioritize simplicity and clarity. Avoid overwhelming users with too much information at once. Prioritize the most relevant data based on the user's current need and context. For instance, someone planning a trip across the city only needs an overview of major transport options and their estimated travel times, while someone waiting at a specific bus stop needs real-time updates on their bus's arrival. The visual presentation should adapt to provide this specific information depending on the user's location and needs.

Interactive maps are an incredibly powerful tool. They allow passengers to actively engage with the data, exploring different options and customizing their views. Consider implementing zooming capabilities to allow users to focus on specific areas, as well as filtering options to show only the transport modes relevant to their travel plans. Color-coding can also significantly improve usability, with different colors representing different modes of transport, levels of congestion, or service disruptions. Remember, these interactive elements significantly enhance the user experience, offering a dynamic rather than a static representation of transportation information.

Choosing the right chart type is crucial. Different chart types excel in communicating different types of data. Bar charts are ideal for comparing the performance of different transport routes, while line charts are useful for demonstrating trends in traffic flow over time. Pie charts can effectively show the proportions of passengers using different transport modes. However, avoid using overly complex charts that may confuse users or obscure important information. Remember, the goal is clear and immediate understanding, not to showcase technical expertise. Simplicity and clarity are always superior.

Accessibility should be a central concern in data visualization. Consider users with visual impairments. Ensure that all visualizations are accessible through screen readers and other assistive technologies. Use appropriate color contrasts to make the information easily readable for those with low vision. Provide alternative text descriptions for charts and images to enhance accessibility for all users. Creating an inclusive system will benefit everyone and ensure an equitable transport experience.

Consider the use of infographics. Infographics combine data visualization with explanatory text and images, effectively communicating complex information in an engaging and easily understandable manner. A concise infographic depicting the average travel times across various routes during rush hour, for instance, can powerfully illustrate the effects of traffic congestion. This is far more effective than presenting the same data in a purely numerical format. Furthermore, infographics can be easily shared on social media and other platforms, promoting wider dissemination of critical transportation information.

Remember to regularly evaluate the effectiveness of your visualizations. Use analytics tools to track user engagement and identify areas for improvement. Collect user feedback and observe user behavior to gain insights into usability. By routinely assessing performance and incorporating user input, you can iteratively refine your visualizations, continuously enhancing their effectiveness and improving the overall passenger experience. This iterative process ensures the visual presentation remains user-friendly and serves its intended purpose effectively. Con-

sistent monitoring and refinement are key to maximizing effectiveness. Prioritize the user's ability to obtain the relevant information quickly and easily.
Finally, seamless integration of data visualization across various platforms is vital. Ensure consistent presentation across different applications, including mobile apps, websites, and in-station displays. Maintain a uniform design language and visual style to create a cohesive and user-friendly experience. This consistency across all platforms significantly improves the user's overall satisfaction and comprehension of the information provided, regardless of the device or access point used. In essence, a well-integrated system avoids user confusion and frustration and maximizes the impact of your data visualization efforts.

3.5 System Integration & Interoperability

Seamless data exchange is the cornerstone of a truly effective smart city transportation system. Without it, individual components operate in silos, hindering the potential for optimized traffic flow and accurate passenger information. Imagine a scenario where real-time traffic data from sensors isn't shared with the adaptive traffic control system; congestion would persist, and journey times would remain unnecessarily long. This lack of interoperability represents a significant missed opportunity.

Effective system integration requires careful planning and a well-defined architecture. A crucial step is establishing standardized data formats and communication protocols. This ensures that different systems, developed by various vendors perhaps, can "talk" to each other without complications. Adopting open standards, such as those promoted by organizations like the Open Geospatial Consortium (OGC), significantly facilitates interoperability, preventing vendor lock-in and fostering a more robust, adaptable system.

Consider the implications for passenger information. A passenger using a multimodal journey planning app needs access to real-time data from various sources – bus schedules, train arrival times, and even real-time traffic conditions. If these data sources aren't integrated, the app provides incomplete or inaccurate information, leading to frustration and unreliable journeys. A unified data platform, acting as a central hub, aggregates and processes data from different sources, ensuring consistency and accuracy. This platform needs to be robust, secure, and scalable to accommodate the ever-increasing volume of data.

One of the major hurdles in achieving effective system integration is data security. Protecting the privacy of passenger data is paramount. This necessitates implementing stringent security measures, including encryption, access controls, and anonymization techniques. Compliance with relevant data protection regulations, such as

GDPR, is also crucial. Furthermore, the system architecture must be designed to withstand cyberattacks, ensuring the continuous availability and reliability of the transportation system.

The success of system integration also depends on strong collaboration between stakeholders. This includes city authorities, transport operators, technology providers, and even end-users. Regular communication and coordinated planning are essential for ensuring interoperability. Establishing clear roles and responsibilities is key; for instance, defining who is responsible for data quality, system maintenance, and security updates. This collaborative approach also necessitates agreeing on performance metrics for measuring the effectiveness of the integrated system.

Successful integration goes beyond merely connecting different systems; it's about ensuring that data flows smoothly and accurately between them. This requires meticulous testing and validation of the interfaces between different components. Regular system monitoring and performance evaluation are also critical for identifying and addressing potential issues promptly. Using real-time dashboards and performance indicators will allow for proactive identification of problems before they impact service quality.

Ultimately, the goal of system integration and interoperability is to create a unified, efficient, and user-friendly transportation system. This allows for better traffic management, improved passenger information, and ultimately enhances the overall quality of life in the smart city. The benefits extend beyond passenger convenience; optimized traffic flow reduces congestion, emissions, and fuel consumption, contributing to a more sustainable and environmentally friendly urban environment. Achieving this level of integration requires a long-term commitment to collaborative planning, standardized data formats, robust security measures, and rigorous testing and monitoring. The payoff, however, is a truly smart city transportation system that operates efficiently and effectively, serving the needs of its citizens. Investing in robust, well-designed, and secure systems today is a strategic choice to lay the foundation for more efficient, sustainable, and integrated transport networks of the future. The creation of open APIs and documentation is a crucial element of this vision, allowing third-party developers to leverage this data for innovation and enhanced service provision, fostering a vibrant and responsive ecosystem. Investing in training and upskilling of staff is also vital, especially to ensure that they are equipped with the skills needed to effectively manage and maintain the sophisticated technology underlying the system. Through continuous improvement and adaptation, we can create increasingly sophisticated systems which meet and adapt to future challenges and technological developments. The focus should remain on the ultimate goal: creating a transportation system that truly serves the needs of the city and its citizens, today and in the future.

4 Data Analytics for Transportation

Understanding historical traffic patterns is crucial. Analyzing past data, such as hourly traffic counts from specific locations over several years, reveals recurring congestion points and peak travel times. This baseline data forms the foundation for effective predictive modeling. For instance, you might discover that rush hour congestion on a particular highway consistently lasts longer on Mondays than on other weekdays. This insight allows for targeted interventions like adjusting traffic signal timings or deploying extra traffic officers. Remember to consider seasonal variations and special events, refining your analysis to account for these influential factors. The more granular your data, the more accurate your predictions will be. Think about integrating weather data – rain, for example, significantly impacts traffic flow, creating predictable slowdowns. Accessing this wealth of historical data, readily available through city transportation departments or public APIs, will give you a solid understanding of your city's transport ecosystem.

Effective predictive modeling requires more than just historical data; it needs sophisticated analytical techniques. Time series analysis is a powerful tool that helps you identify trends and patterns in traffic data over time, forecasting future traffic volumes with a reasonable degree of accuracy. Machine learning algorithms, specifically those used in regression analysis, are particularly well-suited to this task. These algorithms can identify complex relationships between variables, allowing you to anticipate traffic flow based on a multitude of factors, including time of day, day of the week, weather conditions, and even special events. Don't underestimate the power of combining different analytical techniques – a multifaceted approach will produce the most robust and reliable predictions. Experimentation and continuous model refinement are key to improving the accuracy of your forecasts.

Moving beyond prediction, accurate performance measurement is essential for assessing the efficacy of your smart city transportation initiatives. Key performance indicators (KPIs) are invaluable here. Examples include average travel times, vehicle speeds, congestion levels, and the number of traffic incidents. You can use these to track the effectiveness of interventions such as adaptive traffic signal systems or improved public transportation. However, focusing solely on these broad metrics is insufficient; delve into more granular KPIs. For example, instead of just looking at average travel time across an entire city, analyze average travel times for specific routes during specific times. You could even break it down to assess different modes of transportation. Compare pre-intervention data with post-intervention data to quantify the success of your initiatives. This kind of detailed data analysis will provide compelling evidence of the impact of your programs.

Visualizing your data is more than just creating pretty charts; it's about communicating insights clearly and effectively. Dashboards presenting real-time traffic conditions are incredibly useful, giving you an immediate overview of the transportation network. Interactive maps highlighting congestion hotspots, average speeds, and incident locations can help you quickly identify problem areas. These dashboards should be easily accessible to transportation managers, enabling proactive response to emerging issues. But consider the broader audience; creating easily understandable visualizations for the general public is equally critical. Well-designed maps displayed on public information screens in transit hubs or on a dedicated website can give citizens real-time travel information and a better understanding of transportation dynamics in their city.

The data you collect and analyze must be reliable. Data quality is paramount, and ensuring accuracy requires careful consideration of your data sources. Are your sensors correctly calibrated? Are your data collection methods robust and consistent? Identifying and addressing data errors is crucial to avoiding flawed conclusions. Regular audits and quality checks are essential steps. Employing data validation techniques, such as checking for outliers or inconsistencies, can help maintain data integrity. When you have access to multiple data sources, comparing and cross-referencing these sources can aid in identifying any discrepancies. This multi-pronged approach will give you much greater confidence in the accuracy of your analysis, and eventually in the decisions informed by it. Ultimately, the value of your data-driven insights rests on the soundness of your source material.

4.1 Predictive Modeling

Predictive modeling offers a powerful way to anticipate future traffic patterns and optimize transportation systems. By leveraging historical data, current conditions, and even weather forecasts, we can create models that forecast congestion, predict demand for specific routes and modes of transport, and even anticipate potential incidents before they occur. This proactive approach allows for timely interventions, minimizing disruptions and improving overall efficiency.

Accurate predictions require a robust data foundation. This includes comprehensive traffic data from various sources like loop detectors, cameras, GPS trackers in vehicles, and even social media feeds that can highlight unusual events or congestion. Data cleansing and preprocessing are crucial to ensure data accuracy and reliability. Missing data needs to be handled carefully, perhaps through imputation techniques, to prevent biases in the predictive model. Moreover, the selection of relevant variables is critical for model accuracy; irrelevant variables can lead to overfitting and poor predictive performance.

Several statistical and machine learning techniques can be applied to build predictive models. Simple time series analysis can be effective for short-term forecasting, capturing daily or weekly traffic patterns. More sophisticated models, such as ARIMA (Autoregressive Integrated Moving Average) models, can account for seasonal variations and trends. For more complex scenarios involving multiple factors, machine learning algorithms like neural networks or support vector machines offer greater flexibility and potential for improved accuracy. The choice depends on the specific application and available data.

Model evaluation is crucial to ensure reliability. Metrics like Mean Absolute Error (MAE), Root Mean Squared Error (RMSE), and R-squared are commonly used to assess the accuracy of predictions. It's important to remember that no model is perfect, and validation on unseen data is essential to avoid overfitting, ensuring that the model generalizes well to real-world conditions. Regular monitoring and recalibration of the model are vital to maintain accuracy over time, as traffic patterns and other influencing factors can change significantly.

Implementing predictive models effectively requires careful integration into existing transportation systems. Real-time data feeds should be seamlessly integrated into the model to ensure that predictions are up-to-date. The output of the model needs to be readily accessible to traffic managers, allowing for timely interventions. This might involve automated adjustments to traffic signals, rerouting of vehicles, or alerts to drivers about potential congestion. The system needs to be designed to handle various scenarios and provide clear, actionable information.

The ultimate goal is to enhance decision-making and resource allocation. Predictive models help anticipate resource needs, such as the number of transit vehicles required during peak hours, or the allocation of emergency services personnel based on predicted incident locations and severity. By accurately forecasting demand, transportation authorities can optimize resource deployment, reducing costs and improving service quality. Moreover, the insights gained from these models can inform long-term planning decisions, such as the expansion of road networks or the development of new public transportation routes.

Beyond traffic flow, predictive modeling can extend to other aspects of transportation. Predicting passenger demand for different modes of transport can optimize schedules and improve efficiency. Predictive maintenance of transportation infrastructure, based on sensor data and usage patterns, can minimize disruptions due to unexpected breakdowns. Likewise, integrating weather data allows us to anticipate the impact of adverse weather conditions on traffic flow, facilitating proactive management strategies. The applications are numerous, all contributing to a more efficient and resilient transportation system. The potential benefits of predictive modeling are substantial. Reduced congestion leads to decreased travel times, lower fuel consumption, and reduced emissions. Improved safety, through proactive incident management, saves lives and reduces costs associated with accidents. Optimized resource allocation leads to cost savings and enhanced service quality. Moreover, the data-driven insights generated by predictive models can inform evidence-based policy decisions, leading to the development of more sustainable and effective transportation systems. It's a move towards a truly smart city.

However, it's important to consider the ethical and privacy implications. The use of personal data in predictive modeling raises concerns about individual privacy. Appropriate data anonymization techniques and strict adherence to data privacy regulations are essential. Transparency in the use of data and the algorithms employed is also crucial for building public trust and ensuring accountability. By addressing these concerns proactively, we can harness the power of predictive modeling while upholding ethical standards.

In conclusion, predictive modeling is an indispensable tool for optimizing smart city transportation systems. By carefully selecting appropriate techniques, validating models rigorously, and integrating them seamlessly into existing infrastructure, we can unlock the potential of data-driven decision-making. This approach leads to increased efficiency, enhanced safety, and improved sustainability in urban transportation, contributing significantly to a better quality of life for city dwellers. The key is to embrace this technology responsibly, balancing the benefits with the ethical considerations inherent in data-driven approaches.

4.2 Performance Measurement

Measuring the success of a smart city's transportation system isn't simply about counting cars or buses. It requires a multifaceted approach, blending quantitative data with qualitative assessments to create a truly comprehensive picture. Effective performance measurement provides crucial feedback, guiding improvements and justifying investments in the long run.

Journey time reliability, a key metric, signifies the consistency of travel times. Frequent, significant deviations from predicted journey times indicate inefficiencies. For instance, consistently lengthy delays at specific intersections suggest the need for traffic signal optimization or perhaps even infrastructure improvements. Analyzing this data can identify bottlenecks and inform decisions about resource allocation. Data visualization tools, like heatmaps showing average journey times across the city's network, make this complex data readily understandable for both technical and non-technical stakeholders.

Beyond journey times, we must consider passenger satisfaction. This isn't solely a matter of speed; it includes factors like comfort, safety, and the accessibility of information. Surveys and feedback mechanisms provide invaluable insights into passenger experiences. Consider, for example, a system lacking clear signage or a user-unfriendly mobile app; these seemingly minor details significantly impact passenger satisfaction, ultimately influencing the success of the entire transportation network. Regular surveys, coupled with analysis of social media comments and reviews, offer a real-time pulse on public sentiment.

A crucial aspect of performance measurement involves economic considerations. This includes analyzing cost-effectiveness of different technologies and infrastructure investments. Comparing the initial investment costs with the long-term benefits, such as reduced congestion, improved fuel efficiency and reduced emissions, is essential for making informed decisions about future projects. For example, comparing the cost of implementing an adaptive traffic control system against the savings achieved through reduced fuel consumption and decreased traffic congestion provides a clear justification for investment. Cost-benefit analyses, carefully considering both short-term and long-term financial implications, are indispensable.

Environmental impact is another critical area of measurement. Smart city transportation systems should aim to reduce greenhouse gas emissions, improve air quality, and minimize noise pollution. This requires monitoring various environmental indicators, such as carbon emissions from vehicles, levels of pollutants in the air, and noise levels at different locations throughout the transportation network. Data analysis tools should track these parameters over time to assess the effectiveness of interventions aimed at mitigating environmental impacts. For instance, a reduction in average vehicle speeds due to effective traffic management can translate to a measurable decrease in emissions. This data, demonstrably showing environmental benefits, is crucial for garnering public support and attracting further investment.

Finally, the equity of access is paramount. A successful smart city transportation system ensures accessibility for all citizens, regardless of their socioeconomic status, age, or physical abilities. This involves assessing factors like the availability of public transport options in different neighborhoods, the convenience of walking and cycling infrastructure, and the ease of using the passenger information system for people with disabilities. Analyzing ridership data disaggregated by demographic groups allows for the identification of inequities in service provision and informs targeted interventions. For example, a lower than expected ridership in a specific area may indicate a need for improved accessibility or better public transport options. Ultimately, a truly smart city transportation system aims for an inclusive and equitable experience for all its inhabitants.

5 Emerging Technologies in Transportation

Autonomous vehicles represent a paradigm shift in transportation. Their potential to drastically reduce accidents due to human error is undeniable; studies consistently show a significant decrease in collisions when comparing autonomous systems to human drivers. The complexities involved in their implementation, however, are substantial. We must consider not only the technological challenges of perfecting autonomous navigation systems and ensuring robust cybersecurity but also the ethical dilemmas surrounding liability in accidents and the societal impact on employment within the transportation sector. Careful regulation and public education are crucial for a successful integration.

Connected car technologies are rapidly expanding, fostering a network of communication between vehicles, infrastructure, and even pedestrians. This Vehicle-to-Everything (V2X) communication allows for real-time data sharing, significantly enhancing safety and efficiency.

Imagine a world where vehicles can warn each other of impending hazards, or where traffic signals adapt dynamically based on the collective movements of connected cars, optimizing traffic flow and reducing congestion. The benefits are clear, but the technical standards and data security protocols must be carefully considered to prevent vulnerabilities and ensure seamless interoperability across different vehicle manufacturers and infrastructure systems. This requires international collaboration and standardized protocols.

Shared mobility services, encompassing ride-sharing, bike-sharing, and scooter-sharing programs, offer a compelling alternative to personal vehicle ownership, particularly in densely populated urban areas. They address parking challenges, reduce traffic congestion, and promote a more sustainable transportation ecosystem. However, their successful integration hinges on careful urban planning and policy development. We need to consider equitable access for all citizens, manage the environmental impact of the services themselves, and ensure the safety and regulation of the often rapidly evolving sharing schemes. Effective partnerships between city governments and private companies are vital. Successful models often involve data-driven approaches to optimize fleet deployment and user experiences, highlighting the increasing role of data analytics in the management of these systems.

The convergence of these technologies presents unique

opportunities and challenges. Autonomous vehicles, for example, could seamlessly integrate with shared mobility platforms, offering fully autonomous ride-hailing services. Connected cars could provide real-time data to inform the management of shared mobility fleets and enhance the efficiency of public transportation. However, the successful integration of these systems requires careful consideration of data privacy, cybersecurity, and ethical considerations. Interoperability between different platforms and systems is also crucial, requiring the development of common data standards and communication protocols. This necessitates not only technological innovation but also significant policy changes and collaborations between stakeholders.

Furthermore, the widespread adoption of these technologies will require significant infrastructure upgrades. The deployment of autonomous vehicles will necessitate the creation of smart city infrastructure that supports their operation, such as high-definition mapping and dedicated communication networks. Connected car technologies rely on advanced communication networks and robust sensor infrastructure. The successful implementation of shared mobility services demands an adequate network of charging stations for electric vehicles and secure docking points for bicycles and scooters. The financial implications of these infrastructural improvements are significant and necessitate long-term strategic planning and investment.

Effective implementation of emerging transportation technologies demands a holistic approach. It is not enough to simply introduce new technologies; we must carefully consider their impact on society, the environment, and the economy. Successful integration requires the active involvement of diverse stakeholders, including city planners, transportation engineers, policymakers, and the public. Engaging in robust public dialogue, ensuring transparency in data use, and prioritizing equitable access to these technologies are vital for building a truly sustainable and inclusive smart city transportation system. The ultimate goal is not just technological advancement, but a more efficient, safe, and sustainable urban environment for all.

5.1 Autonomous Vehicles

Autonomous vehicles represent a significant leap forward in smart city transportation, promising increased efficiency, safety, and sustainability. Their potential impact is transformative, affecting everything from traffic flow management to public transportation planning. However, successful integration requires careful consideration of various factors, from technological readiness to societal acceptance.

The core technology behind autonomous vehicles relies on sophisticated sensor systems, including lidar, radar, and cameras, which work in concert to create a 3D map of the vehicle's surroundings. This data is then processed using advanced algorithms, allowing the vehicle to navigate, make decisions, and respond to changing conditions in real time. Think of it as a highly complex, constantly updating model of the city itself, built and interpreted by the vehicle's onboard computer. The precision required is immense; even a tiny error in perception can have significant consequences.

Developing reliable and robust algorithms is crucial. These algorithms must be able to handle a vast array of scenarios, from simple lane changes to complex intersections with multiple pedestrians and cyclists. Furthermore, they need to be adaptable and learn from experience, improving their performance over time through machine learning techniques. This ongoing learning process is essential for dealing with unforeseen situations and ensuring the safety and efficiency of the system. Robust testing and validation are therefore paramount, demanding rigorous simulations and extensive real-world trials in diverse environments.

The integration of autonomous vehicles into existing transportation networks presents a considerable challenge. Current traffic management systems are not designed to handle large numbers of vehicles operating independently, and significant infrastructure upgrades may be necessary. For example, dedicated lanes or communication networks might be required to facilitate efficient coordination between autonomous and human-driven vehicles. The development of standardized communication protocols and data formats will also be essential for seamless interoperability between different vehicle manufacturers and transportation systems.

Public acceptance of autonomous vehicles is another critical factor. Concerns about safety, reliability, job displacement, and data privacy must be addressed through clear communication, robust safety standards, and transparent data management practices. Public education campaigns can play a vital role in building trust and dispelling misconceptions. Addressing ethical dilemmas concerning accident responsibility and decision-making in critical situations is also crucial for building public confidence.

Beyond the technological and societal aspects, the economic impact of autonomous vehicles must also be considered. Their widespread adoption could significantly alter labor markets, potentially disrupting industries reliant on human drivers. At the same time, they could create new opportunities in areas such as vehicle maintenance, software development, and data analytics. Careful planning and proactive policy interventions will be necessary to mitigate potential negative impacts and maximize the economic benefits of this technological shift. This includes retraining programs for displaced workers and fostering the development of new industries and job sectors.

Finally, the long-term sustainability of autonomous vehicle systems is paramount. The energy consumption of these vehicles needs to be minimized, ideally through the use of electric or other sustainable power sources. The manufacturing process should also consider environmental impacts, focusing on reducing carbon emissions and waste. Careful life-cycle assessment is necessary to ensure these vehicles contribute to, rather than detract from, a city's overall sustainability goals. The entire framework, from manufacturing and deployment to eventual decommissioning, should be considered from a holistic environmental perspective. Only then can we truly harness the potential of autonomous vehicles for the creation of a truly smart and sustainable city.

5.2 Connected Cars

Connected cars represent a significant leap forward in smart city transportation. Their ability to communicate with each other, infrastructure, and even pedestrians creates opportunities for enhanced safety, efficiency, and a more seamless travel experience. This interconnectedness allows for real-time data sharing, providing valuable insights into traffic flow, accident prevention, and even individual driver behavior. This data, in turn, can inform city planning and resource allocation decisions.

Consider the potential for reduced congestion. Connected cars, equipped with GPS and other location-based technologies, can anticipate traffic jams and suggest alternative routes, collectively optimizing traffic flow across the entire city network. Imagine a scenario where thousands of vehicles seamlessly adjust their speed and routes based on real-time information, preventing bottlenecks and significantly reducing commute times. This isn't science fiction; it's a realistic possibility rapidly becoming a reality.

Beyond congestion reduction, connected cars offer substantial safety improvements. Vehicle-to-vehicle (V2V) communication systems can warn drivers of impending collisions, providing crucial seconds that could prevent accidents. Similarly, vehicle-to-infrastructure (V2I) communication allows traffic signals to coordinate with vehicles, potentially reducing the likelihood of dangerous situations. These systems, combined with advanced driver-assistance systems (ADAS), are poised to dramatically decrease the number of traffic accidents.

Furthermore, the data generated by connected vehicles provides invaluable information for urban planners. By analyzing aggregated data on traffic patterns, congestion points, and accident hotspots, municipalities can strategically plan infrastructure improvements. They can optimize traffic signal timings, identify areas requiring improved public transport options, and even better allocate emergency services. The precise location data adds a new layer of accuracy and detail to existing urban planning approaches.

However, the widespread adoption of connected car technology presents several challenges. Data privacy and security are paramount concerns. The sheer volume of data generated by these vehicles necessitates robust security measures to prevent unauthorized access and misuse. Furthermore, ensuring data anonymity while still extracting useful insights requires careful consideration of ethical implications. Addressing these issues through clear regulations and secure infrastructure is vital for public trust and acceptance.

The integration of connected cars into existing transportation systems also poses logistical obstacles. The seamless exchange of information between different vehicle manufacturers, infrastructure providers, and city authorities requires standardization and interoperability. Developing common protocols and data formats is critical to avoid creating fragmented, incompatible systems. Collaborative efforts between public and private sectors are crucial to overcome this.

Finally, the economic implications of widespread connected car adoption must be addressed. The initial cost of equipping vehicles with the necessary technology may be a barrier for some drivers, necessitating government incentives or subsidies. Furthermore, the infrastructure required to support V2I communication needs significant investment. A comprehensive strategy considering economic feasibility and equitable access is vital for ensuring successful implementation.

The potential benefits of connected cars are undeniable: increased safety, improved traffic flow, and better urban planning. Yet, the transition to a fully connected automotive ecosystem necessitates careful planning, robust security protocols, and a collaborative approach involving all stakeholders. Addressing the challenges related to data privacy, system interoperability, and economic viability is crucial to harnessing the true potential of connected cars and achieving truly smart city transportation. This presents a significant challenge, but one that promises to revolutionize urban mobility and make our cities more liveable and efficient. The task requires a comprehensive, strategic, and collaborative approach, balancing technological innovation with ethical considerations and pragmatic planning. The future of transportation is connected, and the potential for positive transformation is immense.

5.3 Shared Mobility Services

Shared mobility services represent a significant shift in urban transportation, moving away from individual car ownership towards shared access to vehicles. This paradigm shift offers numerous benefits, including reduced traffic congestion, decreased parking demands, and a lower environmental impact compared to individually owned vehicles. However, successful implementation requires careful consideration of various factors.

Effective integration with existing public transit systems is crucial. Seamless transitions between shared mobility options and buses, trains, or subways are essential to create a truly multimodal transportation network. Imagine a system where a user can seamlessly switch from a ride-sharing service to a light rail system, all planned and paid for through a single app, significantly improving the overall travel experience. This requires robust data sharing agreements and technological interoperability between different operators.

Regulation and policy play a vital role in shaping the shared mobility landscape. Clear guidelines on licensing, insurance, safety standards, and pricing models are needed to ensure fair competition, protect both riders and drivers, and prevent monopolistic behavior. City planning departments should proactively engage with shared mobility providers to strategically integrate services into the urban fabric, considering factors like equitable access to underserved communities and avoiding potential conflicts with pedestrian and bicycle infrastructure.

Data plays a crucial part in optimizing shared mobility services. Real-time data on vehicle availability, demand patterns, and traffic conditions enable intelligent dispatching algorithms, reducing waiting times and improving overall efficiency. Data analysis can also help city planners understand the impact of shared mobility on traffic flow, parking demand, and overall urban mobility patterns, informing further policy development and infrastructure improvements. This analytical insight enables proactive responses to evolving urban demands.

The social and economic impacts of shared mobility need careful consideration. While reducing car ownership can have environmental benefits, the displacement of traditional taxi services and the potential impact on employment require thorough analysis and mitigation strategies.

Moreover, ensuring equitable access for all segments of the population, including those with disabilities and those in lower-income brackets, is critical for building a truly inclusive and sustainable transportation system. The integration of accessible features and affordable pricing models must be prioritized.

Technological advancements are continuously enhancing shared mobility services. The development of advanced driver-assistance systems and autonomous vehicles promises to further improve safety, efficiency, and accessibility. The integration of intelligent transportation systems (ITS) with shared mobility platforms provides opportunities for real-time traffic management, optimized routing, and improved passenger information. This convergence will redefine the future of urban transportation.

Successful shared mobility initiatives depend heavily on user adoption. User-friendly mobile applications, clear pricing structures, and reliable service quality are essential to build trust and encourage widespread use. Effective marketing and public education campaigns can help promote the benefits of shared mobility and overcome potential barriers to adoption. Promoting the advantages of convenience, cost-effectiveness, and environmental sustainability will be key to increasing usage.

The future of shared mobility hinges on collaboration. Effective partnerships between private companies, government agencies, and research institutions are vital for fostering innovation, addressing challenges, and realizing the full potential of shared mobility in creating smarter, more sustainable cities. Open data initiatives, collaborative platforms for data sharing and systems integration, and joint policy development efforts will accelerate the maturation of this technology. Only through cooperation can we fully unlock its benefits.

Planning for future infrastructure must account for the potential impact of shared mobility. Sufficient charging infrastructure for electric vehicles, designated parking areas for shared vehicles, and integrated multimodal hubs will be crucial for the seamless integration of shared services into the existing urban fabric. Forward-thinking urban planning is vital to meet the requirements of a rapidly changing transportation landscape. The focus must shift to accommodate the shared transport paradigm's unique needs.

Ultimately, the success of shared mobility hinges on its ability to provide a convenient, affordable, and sustainable alternative to private car ownership. By addressing the challenges and harnessing the opportunities, cities can leverage shared mobility to improve traffic flow, reduce emissions, and create a more equitable and efficient transportation system for all residents. Continuous evaluation, adaptation, and refinement of policies, technologies, and services are paramount to achieving the vision of integrated, sustainable urban mobility.

6 Smart City Transportation Planning & Policy

Effective smart city transportation planning necessitates a holistic approach, integrating sustainable strategies with robust policy frameworks. Prioritizing sustainable modes like cycling and public transit through dedicated infrastructure and financial incentives is crucial. This requires careful consideration of land use planning, ensuring that residential, commercial, and recreational areas are well-connected by efficient and accessible transport networks. Ignoring the interconnectedness of urban development and transport systems leads to inefficient and unsustainable solutions. Therefore, a strong collaborative effort between urban planners and transportation specialists is essential.

Successful implementation hinges on strong public-private partnerships (PPPs). These partnerships leverage the expertise and resources of both the public and private sectors. The public sector brings regulatory power, land access, and long-term vision. Private entities contribute innovative technologies, financial capital, and operational efficiency. Well-structured PPPs clarify roles and responsibilities, ensuring accountability and transparency. A key aspect involves clearly defined performance metrics and dispute resolution mechanisms, mitigating potential conflicts. Careful consideration must be given to contract design, ensuring it aligns with public interest goals and avoids overly favorable terms for private partners. For instance, a revenue-sharing model might be more appropriate than a fixed-fee contract, offering incentives for exceeding performance targets. Such an approach ensures that the private sector shares the burden of financial risk. Clear and comprehensive policy and regulatory frameworks are critical. These frameworks should promote innovation, ensuring that new technologies and services can easily integrate with existing infrastructure. They must address data privacy concerns, outlining clear guidelines on data collection, usage, and protection. Additionally, regulations need to incentivize sustainable transport options, discouraging private vehicle use through measures like congestion pricing or parking restrictions. These frameworks should be adaptive, allowing for adjustments as new technologies and challenges emerge. Regular review and updates, combined with a mechanism for public feedback, are crucial for ensuring relevance and responsiveness.

Meaningful stakeholder engagement is paramount. This involves actively consulting with all relevant parties, including residents, businesses, transport operators, and government agencies. Transparent communication channels, such as public forums and online platforms, are essential for ensuring that diverse voices are heard. Incorporating feedback into policy decisions enhances the project's legitimacy, increasing the likelihood of successful implementation. Techniques like participatory mapping and scenario planning can significantly enhance stakeholder involvement, allowing for a collaborative decision-making process. A collaborative approach fosters a sense of ownership amongst community members, thereby ensuring wider acceptance and support for the implemented changes.

Anticipating future trends and challenges is essential for the long-term viability of smart city transportation systems. The increasing prevalence of autonomous vehicles, for example, necessitates adapting infrastructure and regulations to safely accommodate this technology. Addressing the potential for increased traffic congestion due to autonomous vehicles requires proactive measures, such as optimized traffic management strategies and potential regulatory restrictions on autonomous vehicle usage during peak hours. Likewise, ensuring equitable access to smart transportation systems for all social groups, regardless of socioeconomic status or physical abilities, must be a primary consideration, preventing digital divides. Regular assessments of transportation system performance through rigorous data analysis will facilitate proactive mitigation of future challenges, enhancing the overall resilience of the smart city transport system.

6.1 Sustainable Transportation Strategies

The shift towards sustainable transportation is not merely an environmental imperative; it's a critical component of building truly smart cities. We must move beyond simply optimizing existing systems and actively design for a future where emissions are minimized, and accessibility is maximized for all citizens. This requires a multifaceted approach encompassing policy, infrastructure, and technological innovation.

Promoting cycling and walking is crucial. Investing in dedicated, safe, and well-maintained cycling infrastructure, coupled with pedestrian-friendly urban design principles, encourages active transportation. This not only reduces reliance on private vehicles but also contributes to healthier lifestyles and improved urban livability. Careful planning, which considers factors like proximity to amenities and employment hubs, is necessary to ensure that these modes of transport become genuinely viable alternatives for a large segment of the population. Furthermore, integrating cycling and walking routes into broader public transportation networks can significantly enhance accessibility and convenience.

Public transportation needs a significant overhaul. Improving the efficiency and reliability of bus and rail networks is paramount. This requires significant investment in infrastructure upgrades, such as modernizing rolling stock, optimizing route planning based on real-time data, and implementing smart ticketing systems for seamless and user-friendly experiences. Beyond improvements in infrastructure, exploring innovative solutions like electric buses and trams is crucial for environmental sustainability. Furthermore, integrating real-time information systems allows passengers to make informed choices and optimize their journeys, fostering greater reliance on public transport. Careful consideration should also be given to addressing accessibility issues, ensuring equal access for all members of society.

Electric vehicles are a pivotal aspect of achieving sustainable transportation goals. Incentivizing their adoption through subsidies, tax breaks, and the expansion of charging infrastructure is essential. The widespread installation of strategically located charging stations, particularly in residential areas and along major transportation routes, can overcome range anxiety and encourage wider adoption. Further to this, city-wide strategies to integrate electric vehicles into existing traffic management systems must be implemented to maximize their effectiveness. The integration of smart charging systems that optimize energy consumption during off-peak hours also holds considerable promise for efficient grid management.

Beyond electrification, car-sharing programs and ride-pooling services offer significant potential. These schemes reduce the number of vehicles on the road, leading to decreased congestion and emissions. Effective implementation requires sophisticated algorithms for efficient matching of riders and drivers, along with reliable and user-friendly mobile applications. Careful consideration of pricing strategies is essential to ensure affordability and accessibility while maintaining the financial viability of these services. Furthermore, integrating car-sharing schemes into wider public transport networks can enhance their effectiveness and appeal.

Finally, successful implementation of sustainable transportation strategies requires strong political will and robust collaboration between various stakeholders. City governments need to establish clear policy goals, allocate adequate resources, and ensure effective enforcement. Collaboration with private sector companies is essential for innovation and investment in new technologies and infrastructure. Engaging with citizens, particularly through public consultation and educational campaigns, is vital to fostering community support and adoption of sustainable transportation choices. The collective action of policymakers, businesses, and citizens is what will truly enable smart cities to make this crucial shift towards a greener, more efficient, and equitable transportation system for everyone.

6.2 Public-Private Partnerships

Public-private partnerships (PPPs) are crucial for successful smart city transportation projects. Their ability to combine the public sector's knowledge of city needs and regulatory frameworks with the private sector's innovative technologies and financial resources makes them an invaluable tool. However, effective PPPs require careful planning and management to mitigate potential risks and maximize benefits.

A successful PPP begins with a clearly defined project scope and objectives. This involves detailed assessment of the city's transportation challenges, potential solutions, and the expected outcomes. Crucially, this initial stage must establish specific, measurable, achievable, relevant, and time-bound (SMART) goals. Failure to define these parameters clearly can lead to conflicting expectations and project delays. Open communication and collaboration between public and private partners are key throughout this process, fostering mutual understanding and alignment of interests. This collaborative approach significantly impacts the success rate of the partnership.

Financial structuring is another critical aspect. The agreement should clearly outline the financial contributions of each partner, including upfront capital investments, operational costs, and revenue sharing mechanisms. Transparent and equitable financial arrangements are fundamental to building trust and ensuring the long-term viability of the project. Different models exist, including build-operate-transfer (BOT), design-build-finance-operate-maintain (DBFOM), and concession agreements, each with its own advantages and disadvantages depending on the specific project. A thorough cost-benefit analysis should be undertaken to determine the optimal financial model.

Risk allocation is equally important. PPPs inherently involve various risks, including technological, financial, regulatory, and operational risks. A well-structured agreement should carefully allocate these risks between the public and private partners based on their respective expertise and capabilities. Clearly defining risk mitigation strategies, including insurance coverage and contingency plans, is essential to protect both partners from unforeseen events that could jeopardize the project's success. The sharing of responsibilities needs to be meticulously established during the negotiation phase.

Effective governance structures are essential for overseeing the PPP and ensuring accountability. This typically involves a joint steering committee comprising representatives from both the public and private sectors. The committee's role is to monitor progress, resolve disputes, and make decisions on key issues. Regular reporting and performance monitoring mechanisms are crucial for tracking progress against the project's objectives and identifying potential problems early on. Establishing transparent decision-making processes and clear lines of communica-

tion are pivotal to a successful partnership.
Finally, the long-term sustainability of the PPP needs consideration. The agreement should outline the transition plan after the private sector's involvement ends, ensuring the city can maintain and operate the infrastructure effectively. A well-defined exit strategy, including provisions for asset transfer or continued service provision, is crucial to avoid disruption and ensure the long-term benefits of the project. Furthermore, the agreement should contain clauses addressing potential changes in technology and regulations.
In conclusion, while PPPs offer significant opportunities for developing smart city transportation systems, their success hinges on meticulous planning, transparent financial structuring, effective risk management, strong governance, and a clearly defined exit strategy. A collaborative and communicative approach between public and private partners is paramount, ensuring a mutually beneficial outcome and leading to sustainable improvements in urban transportation. Careful consideration of these factors will significantly increase the likelihood of achieving the desired outcome and maximizing the overall benefit for the city.

6.3 Policy & Regulatory Frameworks

Effective smart city transportation hinges not only on technological prowess but also on robust policy and regulatory frameworks. These frameworks provide the essential legal and administrative foundation, guiding the implementation, operation, and ongoing evolution of smart city transportation systems. Without carefully considered policies, even the most sophisticated technologies can fall short of their potential, leading to inefficiency, inequity, and ultimately, failure to meet the needs of the community.

Strong data privacy regulations are crucial. The collection and analysis of vast amounts of transportation data, a cornerstone of smart city initiatives, raise significant privacy concerns. Robust policies must be in place to protect sensitive personal information while allowing for the data-driven optimization of transportation systems. This involves transparent data governance structures, clearly defined data usage protocols, and mechanisms for individual data access and control. These aren't just abstract concepts; they represent the practical safeguards that will ensure public trust and acceptance. Without them, the entire system risks public backlash and even legal challenges.

Furthermore, interoperability standards are paramount. Smart city transportation systems rarely operate in isolation. They interact with various other systems, including traffic management, public transit scheduling, emergency response, and parking management. The absence of interoperability standards can create fragmented systems, hindering data sharing and coordinated operations. Consistent standards facilitate seamless data exchange, improving system efficiency and reducing redundancy. This also reduces costs, both during implementation and ongoing maintenance. Creating a framework for these standards requires collaboration between different government agencies and private sector stakeholders.

Funding mechanisms require careful consideration. Smart city transportation projects often involve substantial upfront investments in infrastructure and technology. Diverse funding models are necessary, encompassing public sector funding from national, regional, and local governments, as well as private sector investment through public-private partnerships. This ensures financial sustainability and the efficient allocation of resources. Clear guidelines on funding allocation, project evaluation, and cost-benefit analysis are essential for transparency and accountability. The absence of such guidelines can lead to misallocation of funds and ultimately to project failure.

Regulatory oversight needs a flexible and adaptable approach. The rapid pace of technological innovation necessitates a regulatory environment that can adapt to evolving technologies and their associated risks. Rigid regulations can stifle innovation, while overly lax regulations can compromise safety and security. A balance is needed, one that encourages innovation while maintaining adequate safeguards. This requires a continuous dialogue between policymakers, technology developers, and community stakeholders, ensuring that regulations remain relevant and effective in the face of rapid technological changes and evolving societal needs. Effective regulatory bodies are also needed to ensure proper oversight and compliance.

Equitable access is a core societal imperative. Smart city transportation systems should strive for equitable access for all members of the community, regardless of socioeconomic status, age, or ability. Policies should address the digital divide and ensure accessibility for individuals with disabilities. Accessibility features are important in the design of smart transportation systems, as they must cater to the needs of the elderly, those with disabilities, and those unfamiliar with technology. The goal is an inclusive system; systems that exclude parts of the population will naturally meet with criticism and resistance.

Finally, ongoing evaluation and adjustment are vital. The effectiveness of smart city transportation systems should be regularly assessed through the collection and analysis of performance data. This ongoing evaluation process should inform policy adjustments and system improvements, ensuring the continuous optimization of transportation services. This evaluation needs to be systematic, using key performance indicators (KPIs) such as traffic congestion levels, travel times, and passenger satisfaction, to ensure objectivity and allow for informed decision-making. Continuous improvement is not an option, but a necessity for maintaining a truly smart transportation system.

6.4 Stakeholder Engagement

Effective stakeholder engagement is crucial for successful smart city transportation projects. Ignoring this vital aspect can lead to project delays, budget overruns, and ultimately, a system that fails to meet the needs of the community it serves. Building consensus and securing buy-in from the outset is paramount. This necessitates a multifaceted approach, carefully tailored to the specific context of your project.

Open communication is the cornerstone of any successful engagement strategy. This isn't just about disseminating information; it's about creating a two-way dialogue. Regular town hall meetings, online forums, and targeted surveys can provide avenues for feedback. Actively soliciting input from residents, businesses, and other stakeholders demonstrates your commitment to their concerns and fosters a sense of shared ownership. Remember to tailor the communication style to the audience, using clear, accessible language and multiple communication channels for optimal reach.

Understanding your stakeholders' needs and perspectives is critical before even starting the process. Who are your key stakeholders? Residents, businesses, commuters, transit agencies, city officials, and even environmental groups all have a stake in the outcome. Each group has unique priorities and concerns. For example, residents might be most concerned about noise pollution or traffic congestion near their homes, while businesses might prioritize efficient freight movement and accessibility. Thoroughly researching each stakeholder group's interests and potential concerns is crucial for crafting a relevant and effective engagement plan.

Effective engagement requires more than just listening; it also demands proactive action. Responding promptly and transparently to concerns, addressing criticisms constructively, and incorporating feedback into project plans builds trust and confidence. If issues are left unresolved or feedback is ignored, this can undermine the entire process. Moreover, showing a willingness to compromise and adapt the project as needed demonstrates a commitment to collaboration and partnership.

Consider establishing a dedicated stakeholder advisory group. This group can provide ongoing feedback and guidance throughout the project lifecycle. By involving stakeholders directly in the decision-making process, you can ensure that the final project aligns with the community's needs and preferences. Regular meetings, clear communication protocols, and documented decision-making processes should help to build a strong sense of partnership and shared responsibility.

Effective engagement extends beyond the initial stages of project planning. Ongoing monitoring and evaluation are essential to ensure that the system is meeting stakeholder expectations and adapting to changing needs. Regular feedback mechanisms, surveys, and performance reports can provide valuable insights into the system's effectiveness and identify areas for improvement. Remember that building trust is an ongoing process that requires continuous effort.

Don't underestimate the importance of building relationships with key influencers within each stakeholder group. These individuals can play a crucial role in disseminating information, addressing concerns, and building support for the project. Nurturing these relationships through regular communication and collaboration can significantly enhance the effectiveness of your engagement efforts. The power of personal connections is often underrated, and investing time in cultivating these is crucial.

Transparency is paramount; open data policies and accessible information are essential. Making project data readily available to the public fosters accountability and builds confidence in the project's legitimacy. In addition, creating easily understandable visualizations and infographics can communicate complex information effectively and enhance public engagement. This accessibility boosts understanding and fosters a positive perception of the project. Ensure this is easy to navigate and understand, not bogged down in technical jargon.

In conclusion, successful stakeholder engagement in smart city transportation projects requires a proactive, multifaceted approach. By prioritizing open communication, understanding stakeholder needs, responding to concerns, and fostering ongoing collaboration, you can ensure that the final system meets the community's needs and leads to a successful outcome. Remember, effective engagement isn't a one-time event; it's an ongoing process that requires continuous effort and commitment. The time invested in carefully and thoroughly engaging stakeholders will ultimately reap significant benefits, creating a more efficient, effective, and accepted transportation system for all.

6.5 Future Trends & Challenges

The seamless integration of various transportation modes remains a significant hurdle. Imagine a scenario where a passenger needs to seamlessly transition from a subway to a bus, then to a ride-sharing service. Currently, this often involves multiple apps, confusing transfers, and potential delays. A truly integrated system, however, would provide a single, unified interface guiding passengers through their entire journey, optimizing routes in real-time, and offering real-time updates on delays or service disruptions across all modes. This requires overcoming significant technological and logistical challenges, including the development of universal data standards and the harmonization of different ticketing and payment systems. Furthermore, the data security and privacy implications of such a comprehensive system must be addressed proactively to build public trust. This is vital, because without public acceptance, smart city initiatives will struggle to deliver

their full potential.
Data security and privacy concerns are paramount. The vast amounts of data collected by smart city transportation systems offer tremendous opportunities for improving efficiency and safety, but they also raise concerns about the potential misuse of personal information. Robust data encryption, anonymization techniques, and transparent data governance policies are essential to protect passenger privacy while still allowing for the effective analysis of transportation data. Furthermore, establishing clear regulatory frameworks and promoting public discourse on these ethical concerns are crucial to fostering public trust in these technologies. Regulations should ensure data transparency, user control over data usage, and strict penalties for data breaches. Ultimately, public trust will be essential for the widespread adoption of smart city technologies.
Funding and resource allocation remain persistent challenges. Implementing smart city transportation systems requires significant upfront investments in infrastructure, technology, and personnel. Securing adequate funding through public-private partnerships is a crucial step, requiring careful planning, transparent contracting procedures, and mechanisms to ensure accountability. Cities must also prioritize investment based on a comprehensive assessment of their transportation needs and existing infrastructure. This needs to consider the unique challenges presented by the specific city context, whether those chal-

lenges be dense urban environments, sprawling suburbs, or limited financial resources. An understanding of the costs and benefits of various technologies and approaches is essential for the successful implementation of smart city transportation initiatives, especially given the financial complexities inherent in public-private partnerships.

The evolving needs of a diverse population require careful consideration. Smart city transportation systems must be designed to be inclusive and accessible to all members of the community, regardless of age, ability, or socioeconomic status. This includes providing multilingual support, accessible interfaces for people with disabilities, and ensuring affordability for low-income residents. Consideration should also be given to the needs of diverse user groups, including tourists, commuters, and delivery services, each having unique transportation needs and preferences. Understanding these diverse needs is critical to creating truly effective and inclusive smart city transportation systems. Failure to address these diverse requirements risks creating a system that benefits some at the expense of others, undermining the core goal of equitable and accessible urban mobility.

Addressing the skills gap in the workforce is crucial. The successful implementation and operation of smart city transportation systems demand a skilled workforce capable of managing complex technologies, analyzing vast datasets, and adapting to rapid technological advancements. Investing in education and training programs to develop the necessary technical skills is vital, addressing the gaps in expertise and ensuring the availability of a skilled workforce to support the smart city transportation systems' long-term sustainability and operation. Educational programs must focus on practical skills training in areas such as data analytics, cybersecurity, and system integration, as well as broader skills in project management, stakeholder engagement, and policy development. Continuous professional development opportunities will be necessary to stay ahead of the constantly evolving technology landscape. A well-trained workforce is essential for the effective management and continuous improvement of smart city transportation systems, ensuring their long-term success and positive impact on urban mobility.

7 Case Studies & Best Practices

Singapore's success with its Intelligent Transportation System (ITS) offers a compelling case study. Their comprehensive approach, integrating various technologies, resulted in significantly reduced congestion and improved public transport efficiency. Real-time data feeds from sensors embedded throughout the city's road network dynamically adjust traffic signal timings, optimizing traffic flow based on current conditions. This system isn't static; it constantly learns and adapts, using advanced algorithms to predict congestion hotspots and proactively mitigate potential bottlenecks. The integration of this system with public transport scheduling, via real-time passenger information displayed on mobile apps and at bus stops, further enhances the efficiency of the entire transport network. This holistic strategy, incorporating data-driven decision-making, public-private partnerships, and a strong commitment to technological advancement, provides a valuable template for other cities aiming to improve their transportation infrastructure. The success in Singapore demonstrates the importance of a long-term vision, strategic planning, and consistent investment in both infrastructure and sophisticated software. The key takeaway is the synergy between different technological components and their impact on overall efficiency.

Contrastingly, Copenhagen's approach to passenger information systems offers a unique perspective. Instead of focusing solely on a centralized, technologically driven system, Copenhagen emphasizes user-friendliness and accessibility. Their system, while leveraging real-time data, prioritizes simplicity and intuitive design. Information is presented clearly and concisely across multiple platforms—mobile apps, websites, and physical signage at transport hubs. The city focuses on providing information relevant to individual needs, integrating diverse transport modes such as bicycles, buses, trains, and ferries into a single, easily navigable system. Copenhagen's success lies not just in technological sophistication, but in its commitment to user-centric design, ensuring the system effectively serves the diverse needs of its citizens and visitors. By prioritizing simplicity and accessibility, Copenhagen offers a valuable counterpoint to more technologically complex solutions, showcasing that effective passenger information doesn't necessarily require the most advanced technology. The focus on usability, intuitive design, and multi-modal integration delivers tangible benefits, demonstrating that simplicity can be just as effective as complexity.

One significant aspect frequently overlooked is the crucial role of stakeholder engagement. Successful smart city transportation initiatives aren't solely about technology; they necessitate a comprehensive approach incorporating the perspectives and needs of all stakeholders. This includes citizens, businesses, transport operators, and government agencies. Involving these stakeholders from the planning stage through to implementation is paramount to ensure that the system effectively addresses the specific needs of the community it serves. Open communication channels, regular feedback mechanisms, and opportunities for collaborative decision-making are key elements in achieving broad-based buy-in and a successful outcome. Ignoring stakeholder perspectives often leads to systems that fail to meet the needs of the community, resulting in low adoption rates and a failure to achieve the intended benefits. Therefore, effective communication and collaboration are essential throughout the project lifecycle, from initial concept development and feasibility studies to system deployment and ongoing monitoring. The emphasis should be on creating a shared vision and a sense of ownership among all stakeholders involved in the process.

Effective data analysis is critical for optimizing smart city transportation systems. Real-time data streams from various sources—traffic sensors, GPS trackers, smart cards, and mobile apps—provide a wealth of information about transportation patterns, passenger demand, and system performance. The ability to analyze this data effectively is crucial for identifying trends, predicting potential problems, and making data-driven decisions. Advanced analytical techniques, including machine learning and predictive modeling, can be leveraged to anticipate congestion, optimize traffic signal timings, and improve the efficiency of public transport scheduling. However, the success of these analytical methods relies on the availability of high-quality, accurate, and reliable data. Data governance and quality control are therefore vital aspects of any smart city transportation system. Failure to adequately address these aspects can lead to inaccurate insights, ineffective decision-making, and ultimately, a failure to realize the full potential of the system.

Finally, the long-term sustainability of any smart city transportation system requires a comprehensive approach to planning and policy. This entails incorporating environmental considerations, ensuring the financial viability of the system, and developing robust regulatory frameworks to guide its operation and development. Sustainable transportation strategies, such as promoting cycling and walking, integrating electric vehicles, and optimizing public transport, are critical for reducing the environmental impact of urban transportation. Moreover, securing long-term financial sustainability is essential for the continued operation and improvement of the system. This often involves public-private partnerships, innovative funding mechanisms, and efficient resource management. A well-defined regulatory framework is essential for ensuring the safe and efficient operation of the system, protecting user data, and addressing potential challenges and risks. By carefully considering these long-term perspectives, cities can lay the foundation for a robust and sustainable smart city transportation system capable of meeting the needs of future generations. Ignoring these factors risks creating a system with limited lifespan and impact, undermining the significant investments and effort involved in its creation.

7.1 City A: Successful ITS Implementation

City A, a mid-sized metropolis grappling with escalating traffic congestion, embarked on a transformative journey towards a smarter transportation system. Their proactive approach involved a phased implementation of Intelligent Transportation Systems (ITS), prioritizing data-driven decision-making and citizen engagement. The initial phase focused on upgrading existing infrastructure, integrating advanced sensors into the city's road network to collect real-time traffic data, encompassing vehicle counts, speed, and occupancy levels. This detailed data collection formed the backbone of their subsequent initiatives.

The core of City A's success lay in the adoption of an adaptive traffic control system. This system dynamically adjusted signal timings based on the real-time traffic data, optimizing traffic flow and minimizing delays. Crucially, the system wasn't a rigid, pre-programmed solution; it used machine learning algorithms to continuously learn and adapt to changing traffic patterns, evolving throughout the day and responding to unforeseen events like accidents or road closures. This adaptive capacity proved invaluable, significantly reducing congestion and improving overall travel times.

Their investment extended beyond traffic signal optimization. City A implemented a comprehensive traffic monitoring and analysis platform, visualizing the collected data in an intuitive dashboard accessible to both transportation officials and the public. This transparency fostered trust and accountability, allowing citizens to witness the direct impact of the ITS deployment. The platform provided valuable insights into traffic bottlenecks, identifying areas requiring focused improvement. This data-driven approach allowed for targeted interventions, maximizing the effectiveness of resource allocation.

Beyond the technical improvements, City A prioritized effective incident management. Their system integrated with emergency services, providing real-time updates on accidents or breakdowns. This swift response minimized disruption, improving both traffic flow and public safety. Furthermore, the integration of the system with various communication channels, such as mobile applications and social media platforms, allowed immediate dissemination of traffic advisories and alternative routing suggestions to drivers. Proactive communication proved a critical component of their success, reducing public frustration and improving overall satisfaction.

The project's success wasn't solely dependent on technology; it was built upon strong partnerships and stakeholder engagement. City A actively engaged with various stakeholders, including residents, businesses, and transportation agencies, throughout the implementation process. This collaborative approach ensured that the system met the city's unique needs and that the citizens felt ownership of the project. Regular feedback sessions and public forums were crucial in identifying challenges and refining the system's functionality, ensuring the continuous evolution and improvement of the ITS.

Furthermore, City A's success highlights the crucial role of data analytics in optimizing transportation systems. By analyzing the vast amount of data collected through the ITS, City A was able to identify trends and patterns in traffic flow, allowing for proactive adjustments and preventative measures. For example, they identified recurring congestion during peak hours on certain routes, leading to targeted improvements such as dedicated bus lanes or improved public transport connections. This data-driven approach to decision-making greatly enhanced the efficiency and effectiveness of their transportation system.

In conclusion, City A's journey demonstrates that a successful ITS implementation isn't merely about deploying advanced technology; it's a multifaceted process that requires careful planning, continuous adaptation, and strong stakeholder engagement. Their commitment to data-driven decision-making, proactive incident management, and transparent communication set them apart. The successful integration of various technologies into a cohesive system, coupled with a willingness to adapt and learn from real-world conditions, underscores the principles of a truly smart city transportation system. City A serves as a compelling case study, showcasing how the effective implementation of ITS can drastically improve traffic flow, enhance public safety, and enhance the overall quality of life for its citizens. This approach offers a valuable model for other cities aiming to transform their transportation systems and create more sustainable and efficient urban environments. Their experience underscores the critical role of continuous monitoring, adaptation, and a commitment to collaborative, data-driven decision making.

7.2 City B: Innovative Passenger Info System

City B prioritized a user-centric approach, recognizing that an effective system hinges on intuitive design and accessibility. Their solution involved a multi-layered information architecture, seamlessly integrating real-time data feeds from various sources – bus GPS trackers, train schedules, and even live traffic cameras. This comprehensive data integration eliminated the need for passengers to consult disparate apps or websites, simplifying their journey planning significantly. The system dynamically adjusts information based on user location, offering personalized route suggestions and real-time updates on delays or disruptions. For example, if a bus is delayed, the system automatically alerts passengers who have selected that route as part of their journey, suggesting alternative options.

This focus on personalization extended beyond simple route optimization. The city invested heavily in multilingual support, catering to its diverse population. Information was presented not only in text but also through audio announcements in multiple languages and visual cues like icons and maps, ensuring accessibility for visually impaired individuals. Furthermore, real-time data was translated into accessible formats; for instance, estimated wait times were displayed in both numerical and visual representations, with clearly marked large print options available for the elderly or those with low vision. The system was designed with inclusivity as a core principle, ensuring that all residents could benefit from the improvements.

A key innovation in City B's system was its proactive nature. It didn't simply react to disruptions but actively predicted them, drawing on historical data and advanced algorithms. For instance, if historical data showed a specific intersection experienced significant congestion during rush hour on certain days, the system would automatically issue advance warnings to passengers, allowing them to adjust their travel plans accordingly. This predictive capability minimized delays and improved the overall travel experience for commuters. The effectiveness of the predictive model is regularly evaluated and adjusted; feedback from users and operational data inform ongoing improvements to the algorithms.

The system's user interface was designed with simplicity and clarity in mind, avoiding information overload. A clean, intuitive layout prioritizes the most crucial information, such as real-time vehicle location, estimated arrival times, and potential delays. The interface is easily navigable, even for users unfamiliar with technology. The user feedback mechanism, integral to system design, continuously gathers suggestions for improvements and bug fixes. This continuous feedback loop ensures that the passenger information system constantly adapts to the needs of its users. This iterative development process prioritizes user experience and ensures the system's ongoing relevance.

City B's commitment to sustainability extended to its data infrastructure as well. The system utilizes energy-efficient hardware and software, minimizing its environmental impact. The data center powering the system employs green energy sources wherever possible, aligning with the city's broader sustainability goals. Furthermore, data management practices prioritize efficient storage and processing, reducing the system's overall energy consumption. By integrating sustainability considerations into the system's design and operation, City B demonstrates a commitment to responsible technology use. The environmental impact is regularly assessed and strategies are developed to further reduce consumption and improve sustainability. This holistic approach underscores their dedication to a comprehensive smart city solution.

The success of City B's innovative passenger information system extends beyond technological sophistication. It embodies a commitment to inclusive design, proactive problem-solving, and data-driven decision-making. The system doesn't merely provide information; it facilitates a seamless and user-friendly travel experience for all residents, regardless of their technological proficiency, physical abilities, or linguistic background. This emphasis on inclusivity and user-centric design serves as a model for other cities seeking to enhance their public transportation systems. The city actively shares its knowledge and experiences with other municipalities, actively promoting collaboration and knowledge transfer in smart city development.

Beyond the functional aspects, the system's implementation fostered a sense of community and civic pride. The transparent data-sharing practices and user feedback mechanisms built trust between the city government and its residents. Regular updates and communication about the system's performance and future developments further strengthened this bond. This fosters an environment of collective ownership and engagement in the city's technological advancements, demonstrating the system's success as more than just a technological upgrade but as a catalyst for improved civic engagement. Open data policies, allowing third-party developers to access and utilize the system's data, have spurred innovation within the city's technological ecosystem.

The deployment of this advanced passenger information system, therefore, represents a compelling example of how a comprehensive and user-centric approach to technology can create a significantly improved experience for public transport users, leading to a more efficient, sustainable, and inclusive urban environment. This commitment to a holistic strategy highlights the crucial interplay between technological innovation, inclusive design principles, and effective public engagement.

www.ingramcontent.com/pod-product-compliance
Lightning Source LLC
LaVergne TN
LVHW022009200926
848481LV00007B/2304
* 9 7 9 8 3 4 9 3 2 5 5 7 1 *